EARLY BIRD STORIES

What's the Weather?

An Alien's Guide

Alex Francis

Early Reader

Lerner Publications ◆ Minneapolis

First American edition published in 2020 by Lerner Publishing Group, Inc.

An original concept by Alex Francis

Illustrated by Alex Francis

First published by Maverick Arts Publishing Limited

Licensed Edition
What's the Weather?: An Alien's Guide

Lerner Publications Company
An imprint of Lerner Publishing Group, Inc.
241 First Avenue North
Minneapolis, MN 55401 USA

For reading levels and more information, look up this title at www.lernerbooks.com.

Main body text set in Mikado. Typeface provided by HVD Fonts.

Library of Congress Cataloging-in-Publication Data
The Cataloging-in-Publication Data for *What's the Weather?: An Alien's Guide* is on file at the Library of Congress.
ISBN 978-1-72841-508-6 (lib. bdg.)
ISBN 978-1-72841-534-5 (pbk)
ISBN 978-1-72841-535-2 (eb pdf)

Manufacured in the United States of America
1-48664-49087-1/23/2020

Table of Contents

Introduction 6

Reading the Weather 8

Temperature 10

Rainfall 12

Windy Weather 14

Weather Forecast 16

Bad Weather 18

Natural Forecast 22

Quiz 28

Glossary 30

Index 31

Hello fellow aliens!
What do you want to know
about planet Earth?

INCOMING MESSAGE

Dear Finn and Zeek,

I want to visit Earth, but I'm worried about the weather! There are so many different types, I don't know where to start!

Can you please show me some examples and help me prepare?

From,
Dill
(Planet Drull)

Introduction

Planet Earth has lots of different types of weather.

Sun

Wind

Rain

I like
the rain.

Snow

Reading the Weather

There are scientists called **meteorologists.**

Meteorologists have lots of different ways of finding out about the weather!

They use computers to study the weather. They learn about the **temperature** and **rainfall** in different places.

Temperature

Temperature is how hot or cold it is. One way to read the temperature is with thermometers. When it is hot, the level in the thermometer is high.

When it is cold, the level is low.

Rainfall

On cloudy days, rain can fall from the sky.

When it is cold, the rain can turn to snow.

Meteorologists use **rain gauges** to measure how much rain has fallen.

Windy Weather

On windy days, the clouds move fast.

It can go from sunny to rainy very quickly!

This is a **weather vane.**

It shows which way the wind is blowing.

When the wind blows, the arrow turns to face the way the wind is blowing from.

Weather Forecast

A **weather forecast** tells us what the weather will be in the future. This is based on the information that meteorologists collect.

Symbols are used to show the weather.

Sun

Clouds

Rain

Snow

Fog

Wind

Thunderstorm

Sleet

Bad Weather

A forecast can help people prepare for bad weather. **Storms** are one example of bad weather. Most storms don't cause much damage. But some are very dangerous!

Some storms have **thunder** and **lightning.**
Lightning can cause a lot of damage!

Tornadoes are whirling winds that create a funnel shape.

Hurricanes are strong storms that can damage houses and blow cars off roads.

Sometimes a rain storm causes a bad **flood**!

Natural Forecast

You do not need computers to study the weather. You can look at the shape of the clouds. Clouds that are dark and low could mean rain.

Clouds that are high and white mean good weather.

There are other wonderful ways to read the weather!

Birds sing more when the weather is good. Seagulls come inland when the weather is bad at sea.

A red sky at night can mean good weather. A red sky in the morning can mean bad weather is coming.

MESSAGE SENT

Dear Dill,

There are many different types of weather on planet Earth. Meteorologists study it all the time!

If you watch the weather forecast, you can find out what the weather will be like before it happens. Then you can avoid bad weather or put on the right clothes!

From,
Finn and Zeek

Quiz

1. What are the scientists that study the weather called?
   a) Weather pros
   b) Meteorologists
   c) Forecasters

2. What can you use to measure rainfall?
   a) A rain gauge
   b) A weather vane
   c) A thermometer

3. Does this thermometer show that it is hot or cold?

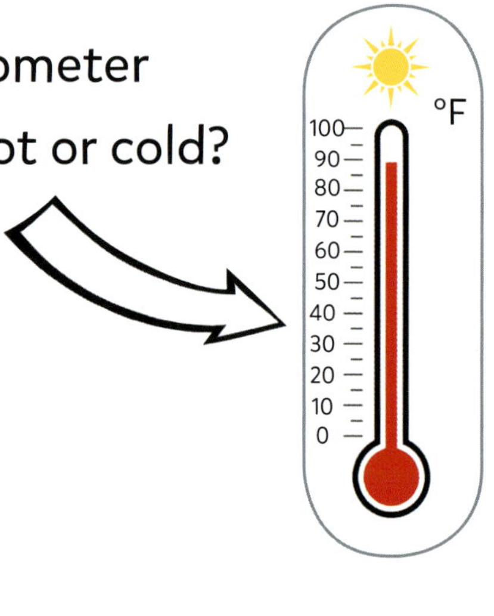

4. What does this symbol mean?

    a) Sunny
    b) Cloudy
    c) Foggy

5. What is the name of a storm with very strong winds?

    a) A hurricane
    b) A mist
    c) A red sky

6. Why are weather forecasts important?

    a) So you know what to eat
    b) It can warn people of bad weather
    c) So you can learn something new

Quiz Answers:
1b, 2a, 3 Hot, 4c, 5a, 6b

Glossary

flood: an overflow of water

hurricanes: storms with very strong winds

lightning: electricity that flashes in the sky during a storm

meteorologists: scientists that read the weather

rainfall: how much rain falls in a place

rain gauges: tools for measuring how much rain has fallen

storms: heavy rain, snow, or hail, often with high winds

temperature: how hot or cold something is

thunder: the loud sound that follows lightning

COLOR	GRL
Silver	L-P
Gold	K-L
Purple	J-K
Orange	H-J
Green	G-I
Blue	E-G
Yellow	C-E
Red	C-D
Pink	A-C

Leveled for Guided Reading

Early Bird Stories have been edited and leveled by leading educational consultants to correspond with guided reading levels. The levels are assigned by taking into account the content, language style, layout, and phonics used in each book.

tornadoes: whirling winds that create a funnel shape

weather forecast: a prediction of what the weather will be

weather vane: a tool for seeing which way the wind is blowing

Index

flood, 21

fog, 17

hurricane, 21

lightning, 19

meteorologists, 8–9, 13, 16

rain, 7, 9, 14, 16, 21–22

rain gauge, 13

storm, 18–19, 21

temperature, 9–10

thunder, 19

weather forecast, 16–18, 20, 23

weather vane, 15

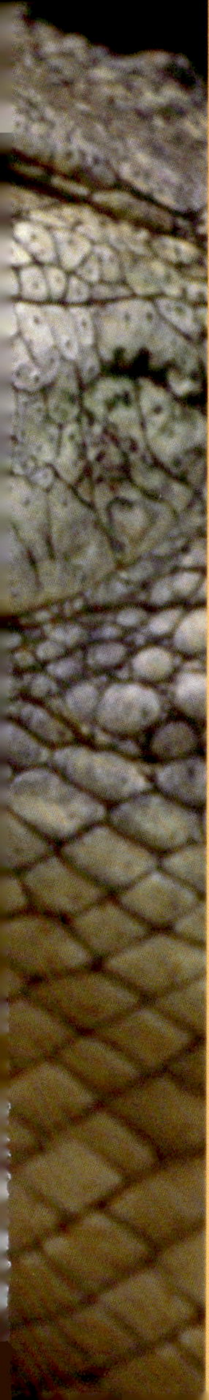

pata

cola

¿Lo encontraste?

escamas

mandíbula

Spot es una publicación Amicus y Amicus Ink
P.O. Box 227, Mankato, MN 56002
www.amicuspublishing.us

Library of Congress Cataloging-in-Publication Data
Names: Klukow, Mary Ellen, author.
Title: Los cocodrilos / by Mary Ellen Klukow.
Other titles: Crocodiles. Spanish
Description: Mankato, Minnesota : Amicus, [2020] | Series: Spot. Animales africanos | Audience: Age 7. | Audience: K to Grade 3. | Includes bibliographical references and index.
Identifiers: LCCN 2018054528 (print) | LCCN 2018055687 (ebook) | ISBN 9781681519029 (eBook) | ISBN 9781681518763 (hardcover : alk. paper)
Subjects: LCSH: Crocodiles–Africa–Juvenile literature.
Classification: LCC QL666.C925 (ebook) | LCC QL666.C925 K5818 2020 (print) | DDC 597.98/2–dc23
LC record available at https://lccn.loc.gov/2018054528

Wendy Dieker y Alissa Thielges, editoras
Deb Miner, diseñador de la serie
Ciara Beitlich, diseñador de libro
Holly Young, investigación fotográfica

Creditos de Imagenes: iStock/vusta cover, 16; iStock/tomodaji 1; Alamy/Colin Marshall, 3; Dreamstime/Arno Meintjes 4-5; Getty/Jason Edwards 6-7; Getty/Mark Deeble and Victoria Stone 8-9; iStock/johan63 10-11; 123RF/sasilsolutions 12-13; Age Fotostock/Roger De La Harpe 14